Fire!

Raging Destruction

by Barbara Bondar

Perfection Learning® CA

Cover Illustration: Stock photo
Inside Illustration: Michael A. Aspengren
Photographs courtesy of Geoff Wilford: pp 10, 27, 29, 32, 33, 34, 44

About the Author

Barbara Bondar is a specialist in interdisciplinary learning and psycholinguistics. She is the author of over 20 highly praised educational books published in the United States and Canada. Her last book for children, *On the Shuttle; eight days in space,* won the 1994 Information Book of the Year Award from the Children's Literature Roundtables.

She has also written and hosted educational television programs and taught film animation to exceptional students. A master teacher, university lecturer, materials reviewer, comic book fiend, and data program analyst, Bondar recently wrote her first interactive computer game for kids of all ages.

She prides herself on being able to start a roaring campfire in pouring rain—a skill her father taught her in northern Ontario and northern Michigan. She has never started a forest fire.

When she is not writing, she offers multimedia presentations to students at all levels on different subjects across the curriculum and motivational presentations to teachers at educational conferences just about anywhere. Check her out at *writebiz.com.*

Perfection Learning® Corporation, 1000 North Second Avenue,
P.O. Box 500, Logan, Iowa 51546-0500.
Phone: 1-800-831-4190 • Fax: 1-800-543-2745
perfectionlearning.com
Paperback ISBN 0-7891-1952-8
Cover Craft® ISBN 0-7807-6114-6
Printed in the U.S.A.
8 9 10 11 12 13 PP 08 07 06 05 04 03

Table of Contents

Chapter 1

Smoke

It wasn't my idea to spend my summer vacation in some dumb forest. It was Mom's. "Do you good to get some fresh air, Mark."

So good-bye, city. Hello, country.

I had to spend a whole summer with my sister Julie. How exciting.

Julie's in her second year of forestry at the local college. She spends her time counting pine needles and studying tiny digs of dirt. I call it dirt. *She* calls it soil. Big deal!

So there we were, hiking back to our campsite near the face of a mountain.

I was looking for UFOs with Julie's binoculars. She was busy counting leaves in squares of *soil.*

And then I spotted it. Smoke! It was the highlight of my day.

Julie took the binoculars and agreed it was smoke. Like it could be anything else.

There was a little catch. There were no phones around. And where was the local fire department when we needed them?

If we'd been in the city, any city, we could have called for help. Or pulled one of those "In case of fire, break glass"

alarms. I may be 11, but I know my way around.

But that wasn't going to work here. Too much forest and purple mountains' majesty between us and anything like a road. Or a phone. Or a proper fire alarm.

The Jule didn't seem worried. She whipped out her map of our area. Then she circled the smoke spot with a pencil.

We headed up the slope on the dusty ski trail we followed. The trees were so tall their leaves kept the sun from us.

And all the while, Julie's telling me not to worry. I think she was a little worried herself. She was walking faster than before.

Why would I worry? It was just a puff of smoke. And a long way off at that.

She explained her plan. First, we'd hike back to our campsite. We'd collect her forest samples and break camp. Then back down the mountain to the Visitor Center. We could phone from there.

Outta here sounded like the best plan to me!

Fuels for Fire

If something is capable of burning, we say it's **combustible.** Then it can be a fuel for fire.

Most everything in a forest is fuel for fire. A forest floor is littered with needles, dead leaves, and bark bits. Of all the forest fuels, this litter is the most combustible.

Julie studies this fuel. She measures off a foot of forest floor. Then she counts.

She counts needles, leaves, twigs, bark, and cones. Anything lying on the surface.

She writes it all up. And the facts go into a computer program about forest fuels.

Julie also tests this fuel for moisture. Here's how she does it.

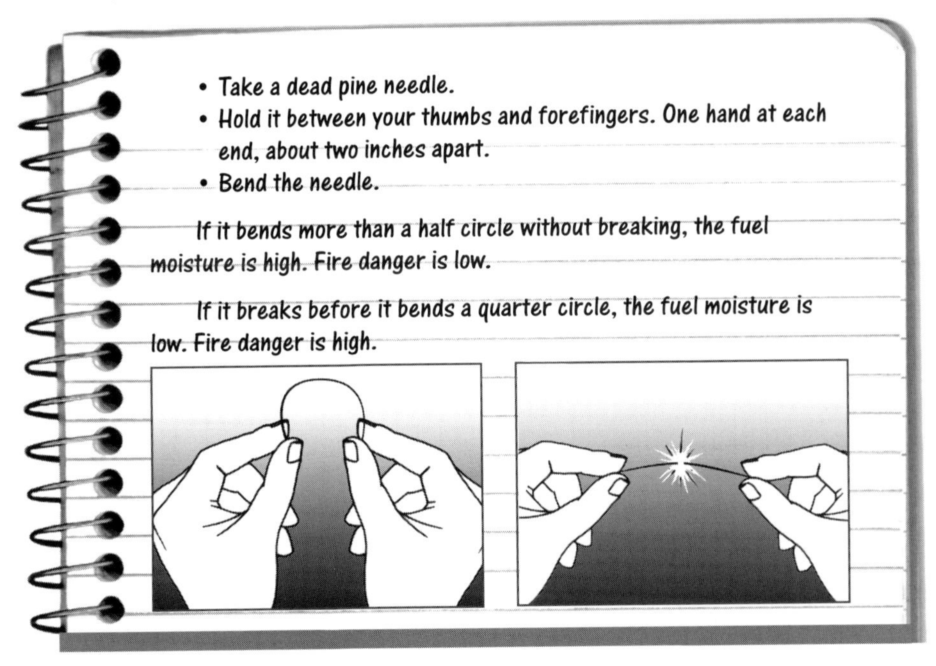

- Take a dead pine needle.
- Hold it between your thumbs and forefingers. One hand at each end, about two inches apart.
- Bend the needle.

If it bends more than a half circle without breaking, the fuel moisture is high. Fire danger is low.

If it breaks before it bends a quarter circle, the fuel moisture is low. Fire danger is high.

Cumulus Clouds

Cumulus clouds are the ones to watch. They mean trouble.

A cumulus cloud is low and builds up vertically. It can grow itself into a cumulus city—a mushroom-shaped pile of cumulus clouds.

The pile has its own range of temperatures right inside it. This cloud crowd can produce ice crystals, lightning, thunder, hail, and strong winds.

The different cloud sections build up plus and minus electrical charges. Big electrical charges mean lightning—bad news over dry forests.

Lightning

Lightning strikes from the cloud edges. Most lightning goes from cloud to cloud. But about 20% travels to the ground. In a

forest, lightning can spiral its way down a tree. It can shatter a rough-barked evergreen. Lightning can ignite a shower of tiny bark embers.

It can scorch right down to the tree root. There it may smolder through decayed vegetation and dry roots. There's trouble if it keeps its heat or hits a pocket with more oxygen. In that case, the underground glow can burn up through to the forest floor.

Old Days	New Ways
You found out lightning struck after a fire broke out.	**Lightning detectors** are wired to computers. The computer records where the lightning strikes. This is printed on a map.

Lightning starts one in three forest fires. The Jule says there are lightning detectors all over North America. If lightning strikes, forest rangers or pilots check for fire.

NOAA

Forest Managers

All North American forests are managed by well-trained people. These forest managers use lightning detectors, satellite photos, and automatic field samplers. They study fire stuff like weather, soil, grass, and forest conditions. All these facts find a place in computer programs.

Fire Danger Ratings

Fire danger is rated by forest managers. Radio, TV, and highway signs announce the fire danger in forests.

Forest managers consider fuel moisture, wind speed, and weather forecasts to make their ratings. They also study facts about the kinds of fuels in their forests. They must know how hot the fuels will burn and how fast they'll spread fire. A lot of facts. A lot of tricky facts.

Forest managers use four words that even we city folks can understand.

> LOW—fires spread slowly and are easily controlled.
>
> MODERATE—fires burn briskly and tend to spread rapidly as they increase in size.
>
> HIGH—fires spread rapidly and may jump to treetops where they spread even faster.
>
> EXTREME—explosive conditions. Fires start easily from sparks and burn fiercely.

Winds

Wind is also a factor. When fire conditions are high, any wind is a threat.

Wind dries fuels and feeds more **oxygen** to a fire. Winds also can carry burning bits of trees. In this way, **spot fires** are started.

It would seem that a thunderstorm would blow out a fire. But no such luck.

When a thunderstorm passes over a fire, it can suck up or push down air. Either way, it increases fire spread.

Chapter 2
Trapped

Julie wanted to save time. So we left the trail. Our escape route was directly through the forest.

We picked our way through a stand of sick-looking trees. Dead branches littered the ground. And the only leaves were a rusty color way above us.

The dead stuff on the ground snapped underfoot. In places, we ducked under fallen trees. Broken branches were in piles as tall as I am.

Forest rangers set up their own fires, or burns, to get rid of this stuff. The fires are set "by prescription." The way a doctor gives you pills.

The rangers start a **prescribed burn.** Then only the diseased area is destroyed.

But they have to wait for the right conditions. Little or no wind. Just the right amount of moisture. Or a rainstorm on its way.

The Jule said these trees had been scheduled for clearing weeks ago. But the weather had been too hot.

This made good sense. Burn out the sick trees and the dead stuff we crunched through. Then new plants can grow back.

New trees. More animals.

We had walked all morning. And we hadn't seen any green leaves since we left the ski trail. What a big hunk of dead trees!

I looked around as we walked. In my mind, I tried to picture a prescribed burn.

Then I smelled smoke. Just a whiff. But there was no mistaking it.

We were surrounded by firewood as far as my eyes could see. So smoke wasn't what I wanted to smell.

I knew Julie smelled it too. I knew because she stopped talking. More like lecturing, really. She had been talking about the trees, the animals…and now forest fires.

Late that afternoon, we finally got to our campsite. I was much happier.

Earlier, we'd pitched our tents in a large, rocky clearing. It was near the top of a mountain.

Below was a huge valley with another mountain in the distance. Way over to the right, lay a sleepy little town. A small river wandered between the mountains.

But now we couldn't see the part to our left. A huge cloud of smoke hung over it.

I helped The Jule pack up. She had her camera and samples of leaves, twigs, bark, and insects.

Let's be honest here—I did the packing. Julie stood for a long time with the binoculars glued to her face. She was studying the landscape.

The wind had picked up. I tucked my T-shirt into my pants. So it wouldn't flap around.

Finally, the Great Jule, Forestry Student and Soil Sampler, spoke. "Mark," she said, "I can't see our down route because of the smoke."

I waited. You could tell there'd be more.

She handed me the binoculars. Much of the mountain we'd just traveled was covered in smoke. I didn't have a good feeling about this.

Julie smiled as she put her arm around my shoulders. "Think we've just been cut off."

"Is that like 'trapped'?" I asked.

She nodded.

The Fire Triangle

In school, the **fire triangle** is easy to draw and understand. Fuel, oxygen, and ignition temperature keep the fire burning. Remove one and the fire dies.

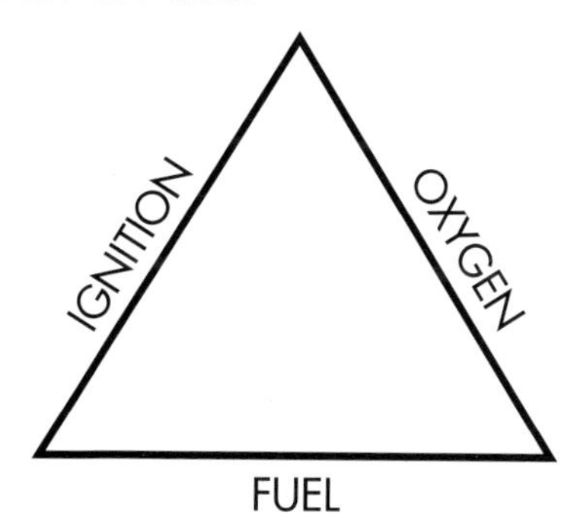

But in a forest, the fire triangle isn't a little drawing with labels. What you realize is that a forest is all fuel. Lying around, standing tall, in your face, and under your feet. As far as you can see. So much for removing fuel.

As for the oxygen, it's blowing at umpteen mph. It blows down the mountains, over the river, and through the woods. So much for removing oxygen.

And miles and miles of fire isn't going to lower the temperature. This is no textbook you can close. So much for lowering the temperature.

This is real. And it's scary!

Bug Thugs

Now, I know it's a bug's job to eat trees. But did you know that insects and diseases kill about seven times more timber than do fires? That's quite a bunch to munch for lunch.

Forest managers use prescribed burns to help stop the spread of disease and pests.

Forms of Forest Fires

The fire that cut us off started as a **ground fire.** It burned under the layer of dry surface litter. It burned through decayed vegetation, buried wood, and old tree roots.

Next it jumped out and became a **surface fire.** It spread, burning all the dead and dry litter on the forest floor. It caught the dead logs and stumps and other brush.

As a surface fire, it climbed up bushes and trees like a ladder. Right up to the top, or crown, of a tree.

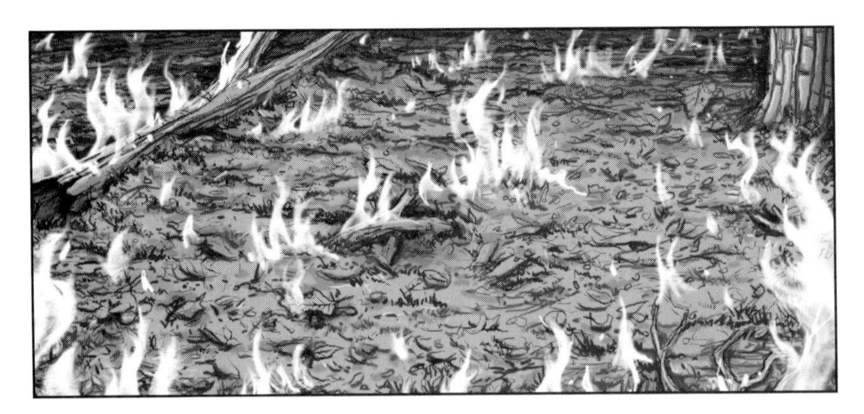

Fire in the crowns of trees can race way ahead of surface fire. Wind helps the flames leap from crown to crown. This is called ***crowning*** or ***topping out.***

The wind can also lift flaming branches and needles from the burning crowns. Then it carries them far ahead to start other fires.

True Understory

Everything that lies under the forest crown is called the ***understory.*** This includes the forest floor litter as well as grasses, shrubs, and short trees.

Forest managers keep close tabs on the understory. Combustible understory is the fuel ladder for fire.

Fires Deserve an Even Break

After a fire, the burned-out area helps slow or stop other fires. This is because the fuel is gone.

Firefighters clear an area, or **firebreak,** to help control the blaze. They dig right down to the soil. They use shovels, bulldozers, and chain saws to remove fuel from this clearing.

Often the fire boss will plan the firebreak next to natural features like swamps, rivers, or lakes. They will slow or stop the fire.

These bodies of water raise the moisture levels around them. Forest fuels are less likely to ignite when the moisture level is high.

If there is no water nearby, the firebreak may be dug beside open fields, rock outcrops, or roads. These provide little combustible fuel to keep the fire going.

Or the fire boss may lay the firebreak next to leafy, or **deciduous,** trees or a stand of hardwood trees like oak or maple. These trees burn slowly. They require greater heat to dry them out before they can be heated to burn.

Smoke Gets in Your Eyes

Smoke occurs because not all the wood burns. Smoke hides what the fire is doing.

Firefighters use special scanners to "see" through the smoke. Some of these scanners are handheld. Some are mounted in the nose of **surveillance** planes that fly over the fire.

The heat from a fire is **infrared** (*infra*—"below" and *red*—"the red limit of our vision"). Even with an infrared scanner, human beings cannot see heat. But the infrared scanner changes the heat image we cannot see into an image we can see.

All living things produce heat. Infrared images are based on this heat.

Each type of tree has its own temperature. So an infrared scanner can detect what kinds of trees are in a forest. It can even show which trees are healthy and which are not!

Old Days	New Ways
People took tree surveys by hiking through the forest. This is how tree types were recorded.	A satellite uses infrared imaging to survey forest trees and discover which of them are not healthy.

Chapter 3

Fire!

Julie decided we'd make our stand against the fire by the rock face. So she set me to work.

Just as well. I felt panicky. Better to keep busy.

We had a **Pulaski** and a collapsible shovel. While Julie chopped at the ground, I started on a trench.

It was tough work. The ground was hard and full of medium-sized stones. Julie told me I had to get down below soil level.

I heard the voice of my basketball coach ringing in my head. He was telling me to concentrate. I did.

I don't know how many hours we worked. But Julie stopped me before I knew I was finished. I'd dug quite a trench.

We placed Julie's samples, the rolled tents, and some other camp gear into the trench. Julie covered it with a solar blanket. Then we piled on as much soil and rock as we could.

When I straightened up, I hurt all over. But Julie stood straight. She had cleared quite a bit too. All the way from the edge of the rock face up to my trench.

But she wasn't complaining. So I wasn't complaining either.

We could see the fire climbing the mountain. It was topping out toward us. Lickety-split it spread through the rusty, dry crowns of that diseased stand of trees.

Bits of burning moss and glowing needles began sprinkling down on us. It was very hot. So was the wind. The smoke that reached us stung my nose and eyes.

Julie ripped a scarf in two and handed me one half. I tied it around my nose like Jesse James. The scarf helped. That's when we heard the airplane.

Julie and I waved towels in the air. But the wind kept the smoke swirling around us. How could the pilot ever see us?

The valley below and to our left was in flames. Black and creamy smoke boiled up toward us.

There was no way the plane could land. It circled us once and flew away.

I couldn't hide my panic any longer. I was too afraid. I knew I was going to be sick. I vomited.

Julie's arm was around me. "That's a surveillance plane."

"What good will that do us?" I asked.

"The **Incident Command** will get to us as fast as they can. We just have to buy some time."

"How do we do that?"

"We do it like some of the forest animals do it. We burrow!" answered Julie.

The Pulaski

The Pulaski is like an ax and a grub hoe. Forest Ranger Edward Pulaski made the first one by hand in 1903.

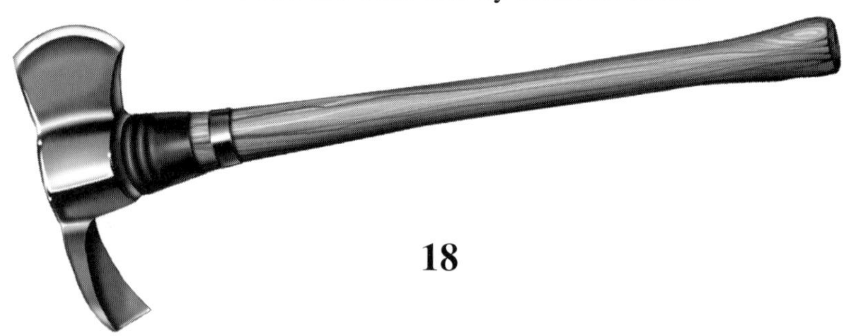

The ax part chops shrubs, roots, and little trees. It also slashes grass.

The hoe part scrapes combustible stuff off the ground. It can dig too. It's a useful tool for pulling the chopped-up fuel out of the firebreak.

How Hot Gets Hotter

The rate at which a fire releases heat is called ***fire intensity.*** Fire intensity increases with larger amounts of burning fuel. The more intense the fire, the faster it heats surrounding fuels.

Here's the perfect fuel for an intense fire.

- a good-sized stand of old, dead trees dried by sun and wind
- an understory piled up with lots of air around it to feed the flames

Just like the diseased stand we walked through earlier in the day. Totally dead and dried and all dressed up for a fire.

Ignition Conditions

Wood glows or smolders between 400°F to 700°F (204°C to 371°C). Wood catches fire, or ignites, at about 800°F (427°C).

Here are some ignition temperatures of stuff around your house.

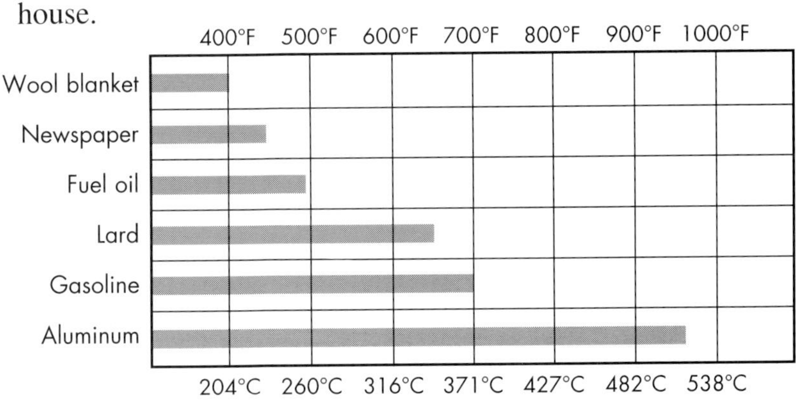

Fire Behavior

Fire spreads out. Wind and slope lengthen its flaming edges into a flattened circle with a burned-out center.

The leading edge of the fire is called the ***head.*** The sides are called ***flanks.*** Wind will blow the head of the fire into new fuel.

The head of the fire moves faster than the flanks. The flames move out more slowly from the already burned-out center.

A fire leaves patches wherever it has traveled. These patches can tell fire stories.

- A stand of deciduous trees won't burn like a stand of conifers.
- Swamp plants won't burn like dry vegetation.
- Night fire doesn't spread as quickly as day fire.
- Slow winds won't speed fire like high winds.

What Animals Do

Some trees have very thick bark that resists surface fires. Many insects burrow into bark. There they wait to resume life after fire passes through.

Some smaller animals like ground squirrels, moles, and field mice shelter deep inside trees or in the ground. Larger animals can move away. Or they can locate themselves in wet areas if they don't panic.

The more intense the fire, the more animals are killed. In large fires, birds may be overcome by smoke, superheated air, or **fire winds.** They fall to their deaths from the sky.

Large fires may produce so much ash that it clogs small water systems. This smothers fish and other water life. As intense fire raises the temperatures in streams, fish may also die of overheating.

Springtime fires kill nesting animals.

Old Days	New Ways
Fire was fought by people who lived near the fire.	Firefighters and their equipment cross borders to put out forest fires and share their knowledge.

Digging In

Julie and I looked like two cowboys in our kerchiefs. We piled the rest of our camp gear on the rock near the overhang.

As we worked, the Jule explained how some forest critters survive. Once in a while, one of us would sneak a peek at the valley below. Nothing but smoke.

Sometimes the smoke would roll away. Then we could see the bright red flames of the fire. That got my attention!

Suddenly, I realized that I couldn't hear Julie. I saw her mouth moving. But I couldn't hear. I had been too busy to notice.

The fire was making so much noise. It sounded like dozens of jet engines roaring at the same time. It was like standing outside on an airport runway.

The heated winds whipped at us like a giant hair dryer. We kept slapping at the burning bits, or ***firebrands,*** that rained down on us. They came from the approaching flames below and behind.

Some of the firebrands were too large to swat. So we danced around to get out of the way.

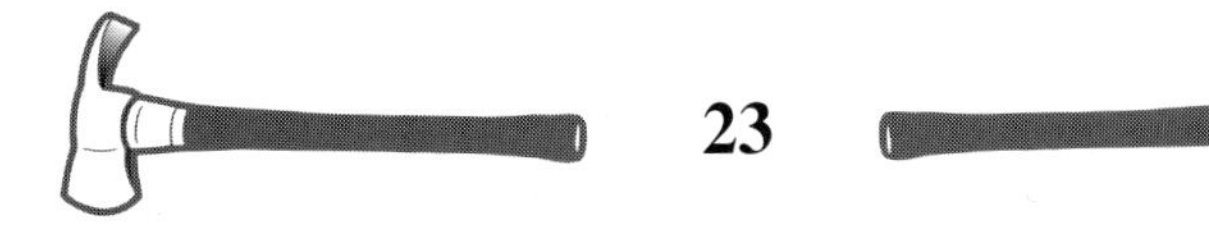

Julie signaled me to come over. She yelled over the noise. "We're going to clear out a burrow under this rock."

We didn't try to talk. We just got to work.

The soil below the rock was part of a large old rock slide. So we were able to move a lot of earth quickly.

My thoughts wandered. What would become of us? I felt Ol' Man Panic stick his nails in my gut.

I whacked harder at the rock slide. Soon we had a good-sized cave beneath the rock. We tucked in our food, water, and other belongings.

The heat was tremendous! The fire roared and snapped. It was like the hot breath of some terrible, hungry beast in a horror movie. Only this was no movie.

Then the branches and trees started flying. No kidding. Not big ones at first. And not too many. But that changed.

Julie kept lecturing. For once I was glad.

She tried to get me to play some "fire games." But I wasn't interested.

We pressed into the cool earth at the back of our burrow. We changed positions often. We couldn't keep one side exposed to the terrific heat for very long.

We kept our faces close to the ground at all times. It was cooler there, and we could breathe better.

I heard the first explosion. Even before Julie spoke, I knew what it was.

One of the boulders above us had become so heated that it had cracked. That got to me. I mean, boulders cracking! I couldn't even think what that meant.

So I got *real* interested in Julie's "fire games" *real* fast.

Home Foam

Before we got into our burrow, Julie added some detergent to water bottles. We shook them and held them upside down with our fingers over the opening. The pressure produced a sloppy spray of foam.

We sprayed foam on the ground in front of the rock. We used it on the cleared soil below us too.

Detergent makes water wetter. The foam gets into logs and soil a little deeper and faster than water. It also doesn't evaporate as quickly. A little detergent and water go a long way.

Old Days	New Ways
People waited for rain or formed bucket brigades.	Foam is delivered by foam generators attached to truck-mounted water tanks.

Julie's "Fire Games"

We tried to think of song titles with the word *fire* in them. I got "Fire's Burning" and "Light My Fire." But Julie's older. So she knew more songs.

We named fire words in different languages. I knew *fuego* and *incendio,* which are Spanish words for *fire.* But Julie knew more.

I knew "if you play with fire, you'll get burned" and "where there's smoke there's fire." Julie didn't know any expressions. Ha!

I knew fireflies and firearms. She knew fire ants.

We did books. We did movies. We did folktales, superstitions, and myths about fire.

She knew some neat stuff. My favorite was an Indian legend from the Lenape tribe of the Eastern Woodlands. It is about Crow.

> To save his animal friends from the cold, Crow flies to the Great Spirit. The Great Spirit gives Crow a stick with fire. It takes three days for Crow to return. The sparks and the soot from the fire have coated Crow's feathers. Crow's voice is hoarse and cracked from the smoke. But Crow has saved his friends.

Convection Direction

Hot air rises from fire. The rising air draws in currents of unheated air. This rising column of air is blown by winds or bounced off mountain slopes. That causes the column of air to circle around inside itself like a chimney. The convection chimney can rise 5 to 7 miles (8 to 11 km) into the air, pulling flames higher.

Firestorms

Violent convection currents grow from intense fire. The superheated chimneys create tornado-like whirls. The winds inside these fire tornadoes can reach speeds of 80 to 100 mph and temperatures of more than 1000°F (538°C).

Firestorms are caused by massive buildups of heat. They are unpredictable. Scientists are still studying them.

A firestorm creates its own weather. Rapidly expanding gases travel in front of the wall of fire. They can snap off trees 60 feet (18 m) above the ground and flatten wooden buildings.

It is not a place for aircraft. So it's not a place for people!

Flame Fame

Flame indicates heat has forced gas from a burning substance. Flame is measured at the front of a fire. It's an indicator of fire intensity.

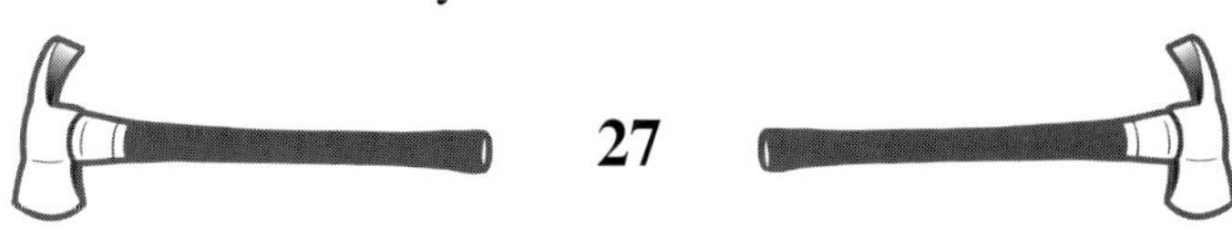

If a flame is

- 4 feet (1 m) or less, fire can be attacked with hand tools
- more than 8 feet (2 m), no control is effective
- more than 11 feet (3 m), fire can spot, jump, and run. If winds are high, the fire will crown.

Firefighters often work in tight spaces. They are careful to aim for the base of a flame. They give short blasts of water with their hoses to the flame base. Then they pause.

A hot fire can turn the water into steam immediately. Steam expands to 1,700 times its original volume of water! Too much water and the steam could scald the firefighters.

Fire Jumping Records

How far can fire jump? Sparks can ride the wind for 1 mile (1.6 km). With help from high temperatures and low humidity, sparks can set spot fires 2 to 3 miles (3–5 km) ahead of the main fire.

During the fire that wiped out Peshtigo, Wisconsin, fire crossed the Straits of Mackinac, Michigan—a jump of 5 miles (8 km)!

Chapter 5

Holding Out

We watched the winds change direction a lot. Julie explained that mountains direct and redirect wind flow. The heat from the fires all around us messed them up some more. The winds circling in the river valley plus the winds from the fire made for a bad burn.

We took turns watching for a break in the smoke. At these breaks, Julie would point out what was happening below us. Sometimes we caught a glimpse of a helicopter towing a bucket beneath it.

Across the valley, firefighters raced against the winds and flames. They hacked away leaves and bushes to dirt level.

There were fewer firebrands now. The roar of the fire lessened.

Once I thought I heard the roar of chain saws. But it might have been a trick of the wind. Or my mind playing tricks on me. Every moment, I was thinking of our rescue.

The firefighters worked all night. I know. I watched them.

In the darkness, we could see the flanks and head of the fire. There were headlights and spotlights. When their light caught the smoke, it looked silver.

The grains and sprinkles and dustings of light worked better than counting sheep. I don't know how long we slept. But the next thing I remember, a bright ray of sunlight poked at my eyelids.

It was morning. There were no more fire sounds above us.

Julie took the map from her backpack. She tossed it outside our earth shelter. It landed on the slope below and flapped in a lazy wind.

She crept toward the opening. She didn't speak to me. She didn't have to. I knew she was testing the temperature.

She stuck her arm out. Then her legs. Then she wriggled from our shelter. Out of my line of sight.

Now I could hear heavy equipment motors. I thought they might be bulldozers. Some chain saws ripped through trees.

Julie shuffled up. She leaned into the burrow. "Come on out, Mark. But be careful what you touch."

Maybe I was getting used to the smoke, but it didn't seem so bad now. My eyes didn't sting either.

We climbed to the top of the rock face. Our hiking boots made loud crunching noises over the scorched ground and burned-out litter.

The flames had passed through. Our mountain world smoldered. It was fever-hot. But it wasn't dead.

Topography

Topography describes the natural features on the earth's surface. Things like mountains and streams and valleys. These features affect a forest fire.

Fire burns more quickly uphill. That's because hot air rises, just like the slope of the hill. So heated air stays closer to the surface than it would on level ground.

Fire will spread almost five times faster going up a slope that is only 50% steeper than level ground. So don't try to outrun a big fire by going uphill.

Narrow valleys and canyons act as natural chimneys. They draw in heat and flames. Fires that burn in these narrow places spread up the slopes faster. They also preheat the fuels on the other side. Stay out of these places in a fire!

Slopes facing the sun dry faster, so they burn more intensely. Water areas increase moisture in the land and air around them. So they slow or stop fire.

Dressed for a Fire

Fire Crew

A fire crew has three to four people. The number depends on the topography of the fire and how much equipment they need to handle. There are two or three firefighters and one crew leader.

The crew leader judges how wide the firebreak should be. And how far in advance of the fire it should be located.

The crews may carry any combination of tools like

- backpack or portable gas-powered pumps
- Pulaskis
- shovels
- chain saws
- radio units
- hoses

Hose Clothes

The chief type of portable hose is made of cloth! No kidding. It's light and can be packed tightly.

You might think that cloth hose would burn up. After all, it has to lie across smoldering fuels. But this all-cloth hose leaks just the right amount of water to prevent the hose from burning. Another Cool Tool.

Old Days	New Ways
Crews got to fires by canoe, railway, handcar, hiking, or pack mule.	Today's air transport gets crews to fire sites quickly.

Helicopters

The helicopter is used for initial fire attack. By dampening flames, it slows the fire advance. A helicopter can douse spot fires and attack the flank of larger fires. It carries crews and equipment to the fire line. The fire boss can also use it as an air-traffic controller or as a "bird dog" to track fire movement.

Helicopters can drop **fire retardant** in small quantities. This is a bright red goop that sticks to everything. It stays cool and cuts off the oxygen. Any goop that gives fire a tough time is a friend of mine.

Rescue

We had our kerchiefs on once again. The wind blew ash over and into everything. What wasn't charred was covered in gray ash.

Stubs of trees stuck up from the ground. They looked like the dirty thumbs of filthy mittens.

Small dust whirls caught ash in funnel shapes. Then the ash dropped like silver baby powder.

The air remained hot. It smelled of smoke, charred trees, and burned grasses. I saw the small bush where I had fed a chipmunk by hand just two days ago. It was a snarled and blackened twist.

We pulled all our stuff out of the burrow. Then we filled in the cave the best we could. We piled everything by the edge of the cliff.

Next we tried to find the stuff we'd buried. The Great Jule had marked each corner with a metal tent peg. We kicked along the ground. I stubbed the first one.

The digging was dirty, hot work. But everything was okay. So we added those things to our pile.

Suddenly, a whistling whine behind us grew high-pitched.

That was all the warning we had. A tree exploded behind Julie. We both fell face forward to the ground.

We were a mess to look at when we stood up. Caked in ash.

Now we knew. Some of the remaining trees and stumps were still smoldering inside.

We rubbed some of the ash off our faces. About that time, we heard a helicopter. Whap, whap, whap. It rose above the valley smoke and noises.

Julie and I jumped up and down to get the pilot's attention. We didn't have to, though. The chopper headed right toward us. Julie and I bent our heads as the vehicle landed.

The helicopter blades made a whirling ash cloud. A firefighter in a smudged yellow jumpsuit appeared out of the cloud.

Without any words, we hugged each other. Then she helped us gather our stuff. We tucked it into stowage space in the copter.

We climbed in. The helicopter lifted off with a gut-dropping lurch. The pilot gave us the thumbs-up. "Ready for more adventure?" he asked.

The firefighter pointed down at the little town. "The forest part of the fire is under control now. But the part that's been burning on each side of the river has jumped a firebreak. And now it's joined up with the last of the mountain fire."

"You need an extra pair of hands? You got 'em," said Julie.

"Actually," said the pilot, looking me over, "we could use two pairs of hands. We've got a town to save here."

Airplanes

Fixed-wing aircraft can carry many crews and more equipment than helicopters. Airplanes used for attacking fire hold larger loads of water or fire retardant than do helicopters. When they are not fire fighting, airplanes are used for surveys,

forest inventories, mapping, and surveillance. Airplanes also make emergency and mercy flights and search and rescue missions.

The wings for fire-fighting aircraft are designed for short takeoff with heavy loads.

Old Days	New Ways
H-boats had open cockpits and could carry little equipment.	While skimming over a water surface, the CL-215 can suck up water through the bottom of its hull—1,441 gallons (5,455 l) in 10 seconds flat.

Famous Firefighters

While we worked, Julie talked. No surprise. Only today it was okay with me. She'd saved our lives. And that makes Julie a hero in my eyes.

She never panicked. Not once. She just tried to use what she had learned in her school and in her reading. I never thought ordinary people could be heroes.

Julie told me about two famous firefighters. They both did ordinary things in extraordinary times.

Edward Pulaski

Edward Pulaski was a career forest ranger. He helped save 42 men in the terrible fire of Coeur d'Alene, Idaho, in 1910. He led the trapped men to a deserted mining tunnel.

Now this was way back in 1910. They had no protective clothing. And they didn't know about convection columns and firestorms.

The heat of this fire was very intense. It sucked the cold air out from the tunnel, replacing it with terrible heat. This left the men gasping for breath. Pulaski made the men lie flat on the ground to breathe.

He used his cap to carry water from a small stream that ran across the mouth of the tunnel. With this water, he fought off the fire in the timbers at the front of the old mine.

Both he and his men were overcome by heat and smoke inhalation. All recovered except five. Pulaski was badly burned.

Red Adair

Red Adair is an oil fire-fighting specialist. On his very first job, Adair remained in an explosive cloud of natural gas. Everyone else ran to escape.

The pressure of the underground gas caused the ground to shake. The vibration loosened the equipment used to prevent an explosion.

Adair remained standing on this wildly shaking ground. Facing the chance of an explosion, he single-handedly tightened every bolt on the well-head valve.

He has improved, redesigned, and invented much of the well equipment used today. He has saved thousands of lives extinguishing fires at gas and oil wells all over the world.

I don't think a person plans to be a hero. I think a hero just does what seems to be right at the moment.

Special Fire Fighting

Every building, every workplace, every industry has special fire-fighting problems. So there are many different solutions.

In schools, there are fire extinguishers, hoses, alarm systems, fire escape routes, fire doors, and hydrants.

Island fires often need fire boats to put them out. Fire boats also carry fire-attack crews to shore.

In some countries where fires rage out of control, the army comes in. The army uses shellfire to set off small avalanches of dirt and rock. These create firebreaks or smother a flank of the fire.

Building materials have to pass certain fire regulations. Buildings are never fireproof. But special materials will slow the fire to give people time to exit.

Enclosed areas like mines and subway tunnels have special problems. These areas suck in air faster when fire starts. Remember, oxygen feeds a fire. So this makes the fire burn faster and stronger.

Checked out an airplane lately? Airplanes are like tunnels. Flames and smoke are quick to spread. Crews are specially trained to help passengers exit quickly. Airlines constantly test and use the latest **fire-resistant** materials.

Tanks are another special group. Think about the big oil tanks that sit in a fueling depot. Or even the gas tank in your car.

When fire threatens these fuel tanks, firefighters first cool the structure if it isn't leaking. This keeps the fluid from expanding and blowing up the tank. These fuels will spread on water. So firefighters use foam to smother leaking tanks.

Most of today's ships have metal hulls. But a fire can force passengers into a cold ocean that can also kill them. Boat drills, built-in fire walls, fire extinguishers, life preservers, and lifeboats are used.

Airports and docks have specially trained crews and equipment. Crews learn to cool heated metal and contain fuel spills that could endanger others.

In the United States, firefighters may carry a fire shelter in their packs. This light-weight, one-person "tent" folds up like a newspaper. Its aluminum coating reflects intense heat.

When trapped by fire, a firefighter can seek temporary shelter beneath the fire shelter. It will withstand more than 600°F (316°C) outside. Inside, the shelter provides the firefighter with a pocket of cooler air. Even inside, the temperature may climb above 130°F (54°C).

Whenever anything is built or invented, fire prevention will always be part of the plan. Whether it is on land, at sea, in the air, or in space.

World-famous Fires

PLACE	DATE	NOTES
London, England	Sept. 2–6, 1666	4 dead 13,000 houses, 85 churches lost 150,000 homeless
Peshtigo, Wisconsin	Oct. 8, 1871	1,500 dead 2,500,000 acres destroyed
Chicago, Illinois	Oct. 8, 1871	300 dead 18,000 buildings lost 90,000+ homeless
General Slocum (ship) New York, New York	June 15, 1904	1,030 dead
Matheson, Ontario	Sept.–Oct. 1916	250 dead 500,000 acres destroyed 6 towns destroyed (At its worst, the head of the fire was 6 miles wide.)
Our Lady of Angels school Chicago, Illinois	Dec. 1, 1958	95 dead
Black Dragon River Valley (border of North China and Russia)	May 1987	220 dead; 250 burned and injured 34,000 homeless 18,000,000 acres destroyed

Chapter 7

Town Fire

My job was to help feed the firefighters. They were covered in charcoal and ash.

There were also dozens of volunteers. They had stayed behind to help save the town after it was evacuated. Many looked tired. But they were excited. They had managed to contain most of the fire.

About ten of us served the food. We kept three long tables loaded—potatoes, eggs, steak. You name it.

Julie was farther up the street. Closer to the actual fire front. She worked on the roof of the little hospital. She had on a bright yellow vest and hard hat.

Her job was to help put out firebrands that fell on the roof. She used wet brooms and mops.

As far as I could see, most roofs had crews with wet mops. The two town-pumper trucks sprayed other roofs and buildings.

The town firemen taught volunteers how to use garden hoses. These worked well on small fires that would break out on dry lawns and in trees.

It was cool! We were all on the same team.

The town firemen also helped organize roof crews. Two or three moved around with walkie-talkies. They reported to the fire **incident commander**. He was set up in a trailer not far from where I worked. Inside the trailer were computers, telephones, a fax machine, maps, and clipboards.

The smoke hung over the river valley. We could hear the snapping rush of superheated air. And it was hot!

But Julie and I were used to it. There had been more smoke and noise and heat when we were stuck on the mountain. I was sure we could win this battle.

Once a fixed-wing aircraft broke through the smoke on the left. It flew across the front of the fire and dropped bright red fire retardant.

There was a great whoosh of air as the fire gasped for oxygen. Its smoke sputtered and fell. The firefighters cheered. A great whoop went up from that part of town.

The fire had reached the last firebreak before the town. Between the fire retardant, the firebreak, and the dampened roofs and buildings, the fire couldn't continue itself. In a few hours, it had burned itself out.

INA Museum

Old Days	New Ways
Hook and ladder trucks were used to carry hats, picks, ladders, axes, buckets, and whale oil torches to fires.	Today we have more equipment, more fire-fighting power, and faster response times.

Computer Fire Fighting

The incident commander computer is a cool tool. It contains fire, weather, and lightning data that is updated every five minutes. It holds local fuel reports and prints maps on demand.

It tracks the whereabouts of firefighters, equipment, and aircraft. It even tracks the food and gasoline that's needed. It has programs to predict where and how fast the fire will spread and how hot it will get. And more!

That's a big reason why only 5% of forest fires grow out of control. A small fire covers a couple of **hectares.** And it's usually beaten within 24 hours.

Fire Talk

Fighting fire is a dangerous, risky operation. So we use a lot of military-type words like

attack	bomb	command
fight	flank	helitack

Timeline of Our Fire

After every fire, reports are made. They add more facts for the fire-fighting computer database.

This is a short report of our fire. This will make a great "How I Spent My Summer Vacation" story!

DATE	SIZE OF BURNED AREA	NOTES
July 17		lightning storm
July 19		fire discovered as surface fire
July 20	280 ft^2 (26 m^2)	spot fires ½ miles (.8 km) ahead
July 20	50 acres (20 hectares)	up-slope burn through diseased trees
July 21	2,200 acres (890 hectares)	high winds merge two valley fires below us; spot fires jump river and race to town
July 21-23		cleanup operations

Feeding the Troops

The fire incident commander ordered food trucked in for the fire crews. The food came with a cook and a portable kitchen.

The United States Forest Service says firefighters should eat 8,000 calories per day. That's four times the normal amount. And that's a lot of food!

The cook gave us our orders. It took five of us to build sandwiches. We stacked canned fruit juice right beside the sandwiches. The firefighters carried this food in their backpacks.

Then on to dinner. We used sturdy power tools to mash the potatoes. With a hand masher, your arm would give out in minutes.

We didn't do anything fancy with the chickens—just cut them in half. The cook did the steaks to order.

We served hot dinners from mid-afternoon until about 2:30 in the morning. The firefighters had been working hard. And we kept their plates loaded.

Mopping Up

The burned area is carefully checked. Surveillance aircraft fly over the scene. Firefighters walk through, extinguishing any glowing embers or logs. They pull down trees that are ready to fall. They report on what needs clearing and **reforesting** and what can be saved.

Fire Where You Live

Believe it or not, there are things you can do to keep fire from where you live.

- Use fire-resistant materials on the roof. This is especially important if your house is made of wood. Slate, asphalt, metal, clay, or fiberglass tiles are fire-resistant. And a sloped roof is best. Then firebrands will roll right off.
- Place any outdoor oil and propane tanks at least 16 feet (5 m) from buildings.
- Make sure you have outdoor water taps. Hosing down your roof will help keep a fire from spreading to your house.
- Keep a well-tended lawn. Then it will be too damp for a fast-burning fire.
- Plan for a wide driveway, rock gardens, or little pools. They all remove fuel from fire.
- Cover chimneys and stovepipes with metal spark arresters.
- Clear leaves, needles, and twigs from roofs and balconies regularly.
- Place fire extinguishers in kitchens and hallways. And learn how to use them.
- Install smoke detectors in hallways outside the kitchen and bedrooms. That's where they'll do the most good.
- Write up some escape plans for your home. Then practice them to see how well they work.
- When you go to school, a movie theater, or a shopping mall, check to see where exits are marked. Figure out how to leave fast.

Chapter 8

One Year Later

You'll never guess where I went the next year. Right back to our "fire" mountain.

City dude or not, I had to check it out.

I knew what happened after city fires. Destruction was fenced off. The new buildings went up.

But what about a forest, two mountains, and a river valley? How do they get rebuilt?

This time I showed up with the right clothes and a new pair of hiking boots. I pinned my old boot laces to my ball cap. We'd been through a lot together. And they still smelled like smoke!

The town had come pretty close to destruction. But it showed no ill effects.

There was still a wide clearing at the edge of town. It had been the last firebreak.

Now it was filled with grass and wildflowers. I even saw the beginnings of a little new building construction.

Along either side of the river, burned-out trees had been pulled down. So they wouldn't fall on hikers or motorists. The river looked clear.

Julie and I hiked up to our former campsite. It was here that the fire had become strong. Charred remains of brush and a few stumps still stood.

The winds of autumn—then winter, then spring—had blown most of the ash from the ground. Some bright wildflowers pushed up through the soil.

When we stood quietly, I could hear a bird singing. Nibbling on its pine cone, a ground squirrel watched us from a charred stump.

The top of the mountain was mostly covered with tall grasses. The whine of chain saws cut through the air. Across the valley, work crews removed trees they could rescue for use.

Our burrow was covered in. The rock slide remained bare. Except for a few weeds. I guess a few weeds is a big deal after a forest fire.

I looked at the weeds more closely. I counted a ladybug, some small flies, and ants in the soil nearby. Something small had nibbled on two of the leaves.

And then it hit me. I was starting to count like Julie!

That night, I thought about my own future. I do that sometimes.

I decided I'd like to have something to do with fire fighting. I remember the panic of being trapped by fire. I remember how it felt to be part of the team that stopped the fire and saved the town.

I liked it. I felt important. And I want to feel that all over again. Often.

How Fires Foster Forests

Fire helps forests and the wildlife that lives in them renew and refresh themselves.

Fire

- prepares sites for seeding and planting.
- reduces forest-fuel hazards.
- helps control disease and insect pests.
- provides new food and cover for animals.
- creates new meadowlands.
- leaves firebreaks through and around forests.
- releases nutrients back into the soil.
- frees the seeds of "fire trees."
- increases the amount of sunlight for seedlings on the forest floor.

Salvage Operations

Salvage operations take place quickly. These happen before burned and charred timber decays or is attacked by insects and disease.

Lumber companies salvage the last of the timber that isn't burned too badly. It is used for products like particleboard and paper.

Recovery Story

Lands do recover on their own. But people can help things along.

Small seedlings from tree farms can be planted along riverbanks. They prevent cave-ins and provide shade to keep fish-spawning areas from overheating. New seedlings can be planted along watersheds as well. This prevents erosion of precious topsoils and stops mud or rock slides. Fast-growing grass seeds coated with fertilizer can be sprayed on topsoil.

Left alone, burned areas become valuable classrooms for us all. Meadowland recovers within one to two years. New plants and shrubs increase rapidly after grown trees are removed.

Forest floors recover in three to five years as seedlings take root in the exposed forest soil. With the grown trees gone, sunlight gets in to aid their growth.

Large animals like elk, moose, and deer return to graze on the tender new growth. With the close-growing trees removed, these large animals can move through the area more easily.

Plants with a Plan!

Not only do trees try to heal their burns, they have other fire plans. The red pine uses its high resin content to seal off injury. This keeps out wood-destroying bacteria and fungus.

Many of the mature pines have thick, corky bark that resists fire unless it is very intense. Lodgepole pines, black spruce, and giant sequoias are helped by fire. The high temperatures melt the resins in their cones and free their seeds.

Others like willow, aspen, and redwood run suckers and sprouts under the newly enriched soil. This starts new growth.

My Future with Fire

Fire is a fact of life on our planet. It occurs everywhere. It can kill and it can cultivate. The best defense is to know what fire needs in order to work. It is also important to understand what it likes in fuels.

But there's so much more to learn. Maybe I'll be the one who discovers how firestorms work. Maybe I can design better, lighter tools for firefighters. Or maybe lighter-weight air tanks. Or clothing that is more heat resistant.

Maybe I can help design a fire-watch satellite. Or the computer programs to make it work. Maybe I can develop better fire equipment for people's homes.

There's so much to do. There's so much to learn. I better get busy. Right now.

GLOSSARY

combustible able to burn

crowning fire that spreads through the tops or crowns of trees. Also called *topping out.*

deciduous trees green, leafy trees that lose their leaves in the fall

fire intensity the rate at which fire releases heat

fire retardant a thick, wet liquid that does not support combustion

fire triangle a model that shows how three things support a fire: fuel, oxygen, and heat. Remove one and the fire dies.

fire wind a tornado-like wind caused by intense fire

fire-resistant contains something that slows the spread of flame

firebrand a flaming bit of forest fuel carried by wind from a crown fire

firebreak an area cleared of forest fuels

flanks the flaming edges at the sides of a moving fire

ground fire fire that glows and burns under the forest floor through decayed vegetation and old tree roots

head the leading edge at the front of a moving fire

hectare 1 hectare = 2.47 acres (10,000 square meters)

Incident Command the control center in a fire-fighting operation

incident commander the fire boss in charge of all who fight a fire

infrared — light energy humans cannot see because it is below (infra) the red limit of our vision

lightning detector — a small equipment station that detects cloud-to-ground lightning strikes. The detector is linked to a computer that records the strikes on a map.

oxygen — a colorless gas that supports combustion

prescribed burn — a fire set by forest rangers to manage land. This fire is used to burn diseased trees or forest litter so new plants grow again.

Pulaski — a combination ax and grub hoe used by fire fighters to clear bush from the path of fire. Forest Ranger Edward Pulaski invented this tool over 90 years ago.

reforesting — replanting an area that was forest

salvage — rescue of damaged but usable wood

spot fire — a small fire started by firebrands falling a few miles ahead of a forest fire

surface fire — fire that spreads out over the forest floor surface, burning dead and dry forest litter

surveillance — close watching or observation

topography — the collection of features that lie on the earth's surface such as mountains, streams, and buildings

topping out — fire that spreads through the tops of trees. Also called *crowning*.

understory — all forest fuels that lie or grow below or under the forest crown

Index